# CRYPTO TAXES
# MADE HAPPY

THE DEFINITIVE
HOW-TO GUIDE FOR PREPARING
CRYPTOCURRENCY TAX RETURNS
IN THE UNITED STATES

MARIO COSTANZ

Happy Tax Publishing
Visit: CryptoTaxPrep.com

# CONTENTS

# DEDICATION

TO JAMES BUDD, my mentor, my business coach, the writer of the Foreword of my first book, the late, the great, the wise James Budd who passed in October 2017 in Chicago, Illinois.

James died young only 60 years old. He didn't drink, didn't smoke, ate healthy and worked out every day. He exemplified someone who cared, lived life to the fullest, took care of others and was selfless. He died two weeks after his life partner Trish died. I think it was of a broken heart so he could be with her again. They were soulmates.

James is someone who I aspire to be more like every day. He taught me how to find balance, how to better handle adversity, acceptance, delegation, letting go, he had a brilliant marketing mind, he was loved by many and he was and always will be my Ben Kenobi.

The time I knew him was brief however it made an indelible impact on who I am now and the path I will travel for the rest of my days. Miss you James. I'm so thankful and happy for the time we had together.

Now, time to go create more victories in your honor James. Thank you for my VisionFire Balance Wheel, Vesuvius, the Pocket of Peace and everything you did for me.

*"It is not the length of life, but the depth."*

– RALPH WALDO EMERSON

# INTRODUCTION

ONCE THEY HIT THE MAINSTREAM IN 2017, virtual currencies quickly became a dominant force in finance. Now, nearly everyone in America knows about Bitcoin and the multitude of alternative coins, or "altcoins," that have followed in its wake. Many of us have heard the stories of the huge sums of money that people have made on cryptocurrencies, and a few of us may be on track to become Bitcoin millionaires ourselves. But relatively few among us – including many tax professionals – have the slightest idea about how the Internal Revenue Service ("IRS") expects us to report and pay taxes on our cryptocurrency income.

The purpose of this book is to provide a thorough overview of the cryptocurrency tax rules that impact the everyday investor. The text begins with a basic primer on cryptocurrency and how it's treated by the IRS. Next, the discussion turns to paying taxes on cryptocurrency earnings, including when you've accepted wages in virtual currency denominations or earned cryptocurrency from hard forks or coin mining. After that, the book focuses on the tax implications of spending, lending, or investing in cryptocurrency. The discussion then shifts to cryptocurrency gifts and inheritances, as well as some helpful tax information for virtual currency retirement savings and life insurance policies. Virtual currency losses and gambling are addressed next, followed by a detailed discussion of different types of virtual currency exchanges. The book closes with an overview of major cryptocurrency-related reporting requirements and an overview of how the IRS has been enforcing its tax policies against the cryptocurrency community.

Before we get into all that, of course, we shall begin at the beginning. Let's get down to the details of what cryptocurrency is and how the IRS has decided to deal with this exciting new technology.

## 1. WHAT IS CRYPTOCURRENCY?

Cryptocurrency – also known as digital currency, crypto, virtual currency, coins, or tokens – is a type of electronic asset intended for use as payments. Despite what the name might suggest, cryptocurrencies are not a new type of currency, at least in the legal sense. Rather, virtual currencies are a new way of exchanging money (and, in some cases, entirely new financial products and services). It doesn't exist in physical form, like the bills and coins of traditional government-issued currencies, but it can be used to purchase goods and services. Some tokens are restricted to certain online communities, while others are increasingly available as retail payment methods in lieu of cash or credit cards.

An anonymous developer (or developers) known only by the pseudonym Satoshi Nakamoto invented virtual currencies a decade ago in 2007, which he announced in a 2008 white paper entitled "Bitcoin: A Peer-to-Peer Electronic Cash System". Virtual currency transactions take place on a blockchain, which is designed to be quicker and more efficient than transactions run through third-party vendors. He shared his invention with the world in the form of Bitcoin, the first-ever application of blockchain technology. Bitcoin's code was made public in its launch, so it was only a matter of time until programmers were able to replicate this technology and develop their own altcoins. Despite increasing competition, Bitcoin still dominates all other cryptocurrencies in total users, network nodes, active addresses, average transaction rate, and average value of transactions. However, more altcoins and hard forks of existing cryptocurrencies are being launched every day. This has given rise to an entirely new financial market that has attracted investors from all around the world.

Cryptocurrencies are designed to allow individuals to use them with near or complete anonymity. This makes them attractive for any number of legitimate applications where financial privacy is important. However, this same quality makes virtual currencies attractive to people engaging in criminal or fraudulent activities, such as tax evasion, money laundering, or the trade of illegal goods. As a result, more and more regulations are being imposed on cryptocurrency markets every day.

## 2. HOW THE IRS DEFINES CRYPTO-CURRENCY

Cryptocurrency is a remarkable invention of the digital age. It is so groundbreaking, in fact, that it does not fit squarely into America's existing set of financial regulations. As a result, the Internal Revenue Service has had to figure out cryptocurrency tax policy as it goes along.

In the United States, for something to qualify as "currency," it must fit the definition of "coin and currency of the United States, or of any other country." Since cryptocurrency does not belong to any country, it cannot be taxed as currency. As a result, the

IRS clarified in Notice 2014-21 that virtual currencies are taxed as property rather than currency. Property is taxed at capital gains tax rates rather than the marginal tax rates applied to other types of income.

For people who purchase virtual currency as an asset or investment, this rule is straightforward. However, it makes things quite difficult for people who use cryptocurrency for other purposes, such as buying or selling goods, making charitable donations, or planning their estate. As a result, the complexity that cryptocurrency users face every tax season may be astronomical, and enforcement issues are equally complicated.

The purpose of this book is not to rehash the history of Bitcoin and its progeny, nor is it to do a deep dive into the benefits and pitfalls of the cryptocurrency revolution. Instead, this text discusses the legal tax implications of buying, selling, using, giving, inheriting, or investing in virtual currencies. So let's get started!

# [1]

# EARNING

# CRYPTOCURRENCY INCOME

LIKE ALMOST EVERY OTHER SOURCE OF INCOME, the IRS taxes income earned in the form of cryptocurrency. However, there are several ways to earn cryptocurrency income. Let's address each one by one.

## 1. CRYPTOCURRENCY WAGES AND FREELANCING INCOME

If you accept cryptocurrency in exchange for goods or services, you must pay taxes on the dollar value of the coins you earned. The value is based on the time of "constructive receipt" of this income, which is the moment when the taxpayer is able to access the funds represented by the cryptocurrency. In other words, since you can cash out your crypto-currency income for U.S. dollars the second you receive them, the IRS taxes them based on their dollar value at that time. Unfortunately, if the dollar value of the cryptocurrency you accepted goes down before you cash out, you are still liable for paying taxes on the coin's value when you received it. As a result, if you accept virtual currency for goods or services, your tax liability vests at the time you accept payment.

Wages paid in virtual currency denominations are taxable. This is a particularly important fact for crypto entrepreneurs to pay attention to, as many new blockchain businesses often compensate early employees using virtual currency. Typically, an employer must report the wages on a Form W-2, just like wage payments in traditional money. Virtual currency wages are also subject to federal income tax withholding and payroll taxes. Employers and employees typically split Social Security and Medicare taxes, with each paying 6.2% and 1.45% for the respective taxes. However, the vast majority of individuals who accept cryptocurrency payments are self-employed. Self-employed individuals must pay the total 12.4% Social Security tax and 2.9% Medicare tax on their own as self-employment taxes, as well as state and federal income taxes.

As with wages, payments to independent contractors in the form of Bitcoin or other cryptocurrencies are subject to the same tax rules most of us are already familiar with. Digital currency payments to independent contractors are taxable as self-employment income, typically filed on Form 1040 Schedule C. The person or business who contracted the work must issue a Form 1099 for any payments of a $600 value or more, and freelancers should report this income on a Form 1099-MISC. However, freelancers are responsible for reporting self-employment income even if they don't get a Form 1099.

Self-employed individuals may not get help from an employer when it comes time to pay employment taxes, but they can deduct ordinary and necessary business expenses from their income tax. Any equipment or software purchased for the business can be deducted, as well as rental payments made on office space. However, as is the case for all deductions, self-employed individuals should be careful to keep all documentation necessary to support their deductions. If not, the IRS may disallow the deductions and charge an accuracy-related penalty.

Virtual currency wages, self-employment income, or other payments should be reported using the full fair-market value of the cryptocurrency at the time the payment was made. So, for example, if you are paid one Bitcoin when the price was $10,000, but the price increased to $12,500 by the time you file your taxes, you report the income as $10,000. Under-reporting your cryptocurrency income or under-paying your cryptocurrency taxes can cause you to rack up penalties and interest from the IRS, so be sure to properly disclose your earnings on your tax return each year.

Since you most likely will not immediately convert the virtual currency you accept for goods or services to US Dollars, you will be subject to an additional tax reporting requirement at the time sell or exchange it. If you accept virtual currency as an employee or a freelancer, be proactive. Discuss your potential tax liability with your tax preparer and strategize for reducing your tax bill. If your regular accountant isn't up to speed on virtual currency, consider working with a specialized cryptocurrency CPA instead.

## 2. CRYPTOCURRENCY MINING IN-COME

Miners are the linchpin of the cryptocurrency ecosystem. Mining bitcoin and other cryptocurrencies is a taxable activity, regardless of whether you do it as a hobby or for a job, if you make more than $400 worth of mined coins for any given year.

Understanding how to properly report your cryptocurrency mining income to the IRS involves understanding your liabilities and deductions. Even if you only mine cryptocurrency on a small scale, the IRS treats mining as a business subject to self-employment tax. Just like driving for Uber or offering guitar lessons, mining cryptocurren-

MARIO COSTANZ | 27

cy is basically a side hustle. As a result, the extent of your tax liability will vary based on several factors.

As long as you aren't an actual at-will employee of a crypto mining business, income from mining qualifies as self-employment business income reported on Schedule C of your annual tax return. The IRS requires miners to recognize income for every coin they mined in a taxable year. When calculating your income, use the market price of each mined coin (or fraction of coin) at the time it was awarded to you. This is the "cost basis" of the coin, which is also used to calculate actual gains or losses in the future if you sell the coin.

By way of example, let's assume that a miner is awarded 1 bitcoin when the market price is $8,000 per BTC. This creates $8,000 worth of taxable income for that year. If this same miner later sells this bitcoin when the price goes up to $10,000 per BTC, she has also generated a taxable gain of $2,000 (sale price minus cost basis). This $2,000 is taxed separately as a capital gain, which is required to be reported on Schedule D of the miner's tax return in the year that she sold the coin.

Fortunately for the crypto miners out there, it's not all bad news. You are entitled to deduct eligible expenses from your self-employment business. Most miners use expensive hardware and tons of electricity to verify transactions on a decentralized block-chain network. These expenses are deductible, as long as your mining operation is substantial enough to be considered an actual business. If your mining activity is not substantial or continuous, your deductions may be impacted.

## 3. HARD FORKS

In the crypto world, a hard fork is when an existing blockchain splits into two (resulting in the original token plus a new one). After a fork occurs, new coins are created and then "airdropped" – meaning that they are given out for free to anyone who owned the original coin (the source of the fork) at the time of the split. Some cryptocurrency investors are thrilled to receive forked coins and the windfall of cash they can bring. However, there is a great deal of uncertainty regarding how the IRS will tax forked coins.

One of the most successful hard forks thus far occurred in August 2017 when Bitcoin Cash forked from the Bitcoin blockchain. This process basically split Bitcoin into two separate coins: Bitcoin ("BTC") and Bitcoin Cash ("BCH"). When this occurred, everyone who held Bitcoin in a compatible wallet or exchange became entitled to claim an equivalent amount of Bitcoin Cash. Anyone who received the coin for free because they held BTC at the time of the fork is celebrating the coin's success. Many investors may not even know that they are entitled to claim the coin, since doing so often requires specific technical expertise. However, BCH can be claimed by any eligible investor that owned BTC at the time of the fork even if they didn't do so at the time of the fork.

Forked coins create an interesting challenge for tax preparation and accounting. In some ways, all forks that you can claim due to your original Bitcoin, Ether, or Litecoin holdings are constructive income, since you can claim them at any time. However, in the multitude of forks that have followed Bitcoin Cash, few (if any) of them created any real value. In fact, some of them were outright scams. As a result, many forked coins will never be claimed out of concern that doing so will expose your wallet to security issues.

So, how does the IRS treat income from forked coins? Unfortunately, there is a lack of consensus in this regard. Some people claim that you must report the forked coin as income when you actually claim it. However, it may not be so simple. Others believe taxpayers should play it safe and assume that every forked coin – claimed or not – qualifies as constructive receipt of income that is taxable at the time the coin forked. Under this theory, cryptocurrency investors would need to report all current and past forks of Bitcoin, Ether, or Litecoin based on the time of the fork and how much of the original cryptocurrency that they held at the time. With new forks being announced every few days now, this can get complicated quickly.

Tax experts are far from reaching a consensus on this question, but most have decided that the IRS will most likely treat cryptocurrency hard forks just like they treat a spinoff of stocks. When a publicly-traded company splits off a portion of its assets into a new company, shareholders get stock in the new company in proportion to the stock they held in the old company. These new shares of stock are taxed under the capital gains rules, with the cost basis typically being allocated between the old and new stock. The gains or losses from the new stock are realized when it is sold, which triggers tax liability.

Given the similarity between hard forks and stock splits, the IRS is likely to treat them the same way for tax purposes. Unfortunately, however, establishing the cost basis of the forked coin can be incredibly difficult due to the volatility that has come to define cryptocurrency markets. Cost basis would need to be calculated based on the value of both the original currency and the new forked coin at the time the forked coin was claimed. Around the time a fork occurs, the original coin usually spikes in value as investors buy in to take advantage of the fork. And forked coins themselves have demonstrated extreme volatility. For example, Bitcoin Diamond – a lesser-known fork of the Bitcoin blockchain – spiked to around $95 immediately after launch before finding its trading range at between $15 and $25. Likewise, Bitcoin Gold, the fork that immediately followed Bitcoin Cash, peaked at around $500 around the time it was launched before slowly declining to a trading range around $100. This volatility shows how difficult it can be to nail down a cost basis, since the value at the time the forked coins are claimed is anyone's guess.

With many forks on the horizon in 2018 and beyond, this can get complicated quickly. Even if you haven't claimed a forked coin you're entitled to, a virtual currency split may be a taxable event in itself. As a result, the safe bet is planning your tax liability by keeping an eye on upcoming forks that will entitle you to airdropped coins. By planning ahead, you can avoid a potentially costly misstep in handling your taxable cryptocurrency assets. If you hold Bitcoin, be sure to consult with a tax professional to make sure you understand how, what, and when you need to disclose information about forked coins to the IRS.

# [2]

# SPENDING OR LENDING CRYPTOCURRENCY

THE IRS CAST A WIDE NET when it established its cryptocurrency tax policy, so pretty much any disposition of virtual currency assets is a taxable event. Tax liability is triggered when you trade your cryptocurrency for cash or other virtual currencies or use it to make a purchase or other type of payment.

This includes paying employees or freelancers, purchasing goods or services, and making loans. Depending on your investment and spending habits, this can make things complicated. The following discussion goes through each of these scenarios and covers the tax issues raised, one by one.

## 1. PAYMENTS TO EMPLOYEES AND FREELANCERS.

Any payment made in cryptocurrency is reportable to the IRS. As discussed previously, employers paying wages in cryptocurrency are held to the same withholding requirements as employers paying wages in U.S. dollars. Additionally, if you paid an individual contractor more than $600 worth of virtual currency, you are required to issue a Form 1099-MISC to that individual and the IRS. Everything you report to the IRS must be broken down to its dollar value, so the amount you must report is the value of the cryptocurrency at the time you made the payment. As a result, if you anticipate paying an individual more than $600 in non-employment wages in particular calendar year, have the person fill out a Form W-9 first so you can properly issue them a 1099-MISC at the end of the year.

Virtual currency payments made to any individual in excess of $600 are subject to backup withholding. In other words, if the person you're paying refuses to give you a social security number or tax ID, you must withhold 28% of the income you paid them and send it to the IRS. Failing to send in backup withholding can expose you to fines and penalties, and even potential criminal liability if the failure was an intentional evasion of taxes.

The IRS requires employers to keep payroll records going back at least three years, and they can go after unpaid taxes retroactively. As a result, good record-keeping is critical. If the IRS inquires about payments you've made using virtual currency, be sure to have documents ready showing the time, amount, and withholdings (if applicable) for any payments you've made to contractors or employees.

## 2. RETAIL PAYMENTS IN CRYPTO-CURRENCY

Contrary to the popular belief – and wishful thinking – of many cryptocurrency investors, retail purchases made using cryptocurrency are reportable and subject to capital gains rules. For example, if you make a purchase using Bitcoin on Overstock.com, this is a transaction subject to capital gains tax if there was a gain on the amount you exchanged the coin for versus the amount you purchased the coin for. This is true no matter how small the purchase is, as the IRS has yet to create any de minimis transaction exemptions. So, even if you get your hands on a virtual currency debit card, you will have to figure out your cost basis and pay capital gains on every purchase you make.

This is a substantial limitation to the use of virtual currencies for retail payments, with very little option to work around it. As a result, most people who use virtual currencies to pay for goods and services first cash out a larger amount for traditional fiat money. Then, they use the dollars exchanged for their crypto coins to make the payments. While this isn't the most convenient option in the short term, it could end up saving you substantial time and money next tax season.

## 3. VIRTUAL CURRENCY LOANS

Under the tax code, lending any type of money is not a taxable event in itself. Only the interest paid on a loan is taxable as income for the lender and potentially deductible by the borrower if it were for business purposes. Theoretically, a loan in cryptocurrency should be treated the same as a loan for traditional cash.

If you are receiving interest on a loan using virtual currency, these interest payments are reportable as taxable income for the payee. As a result, you should be ready for your lender to provide you with a 1099-INT. Even if they do not provide you with one, you are still obligated to report this income on your tax return. This is true regardless of whether the loan was made in cryptocurrency or traditional U.S. dollars.

Take particular care when taking out or accepting loans in the form of cryptocurrency. Doing so may have unintended consequences for you come tax time if the rules evolve in the meantime.

# [3]

# VIRTUAL CURRENCY

# INVESTMENTS

CRYPTOCURRENCIES MAY HAVE STARTED off in the deep corners of the internet, but they have since emerged into the mainstream. Virtual currencies are now a very common investment. Bitcoin futures are even traded on major exchanges like the Chicago Mercantile Exchange and the Chicago Board

Options Exchange. Large investment banks like Goldman Sachs are setting up virtual currency trading desks as well, despite having less-than-friendly cryptocurrency policies on their consumer banking sides. And as cryptocurrency has crept its way into the mainstream financial community, the Internal Revenue Service has refined its policies regarding tax collection and reporting requirements for virtual currency investments.

## 1. TAXES LEVIED ON CRYPTOCURRENCY INVESTMENTS

If you invest in cryptocurrencies, your coins are characterized as a capital asset by the IRS. A capital asset is a piece of property like real estate, vehicles, stocks, bonds, or valuable collectibles like art or antiques. The IRS decided on this definition because virtual currencies are convertible into cash.

Capital assets owned for more than one year produce what is known as a long-term capital gain (or loss). Capital gains are taxed differently than earned wages or self-employment income, and the rates vary based on your tax bracket. With cryptocurrency, capital gains rules apply regardless of how taxpayers decide to use their virtual currencies – whether they are buying and selling for investment purposes or spending coins on goods and services.

When you purchase a cryptocurrency, you've established your cost basis. However, the asset is not actually taxed until you sell it. When you sell it is when you "realize" your gains or losses on the investment. So, if you bought one bitcoin at $12,000 and sell it for $13,000, your realized gain is $1,000, even if it dips below your initial purchase price at some time in between. Sounds simple enough, right? Unfortunately, it gets more complicated.

As is true for all capital assets, your tax rate depends on the length of time holding (or hodl'ing as is the common terminology in the crypto space). If you held a cryptocurrency asset for less than one year before selling

it or swapping it for a different virtual currency, you are taxed at short term capital
gains rates which are equal to your ordinary
income tax rate. This rate varies between 0%
and 39.6% for 2017 and based on how much
taxable income you had in the tax year (the
top individual tax rate is reduced for 2018 to
37%). However, if you hold these assets for
more than one year before selling or trading
them, you will be taxed at the long-term capital gains rate. Long-term capital gains taxes
are lower than short-term capital gains tax.
Long-term capital gains rates in 2017 vary
from 0% for many people to a maximum of
20%. In 2018, if you are married filing jointly, you can have income up to $479,000 be-

fore you are in the top 20% long term capital gains rate.

Capital gains and losses must be reported accurately on your tax returns. If you lost more on your cryptocurrency investments than you made, you can claim up to $3,000 of the capital loss as a deduction on your adjusted gross income in that tax year with any additional loss over that amount carrying to future years to be deducted against any capital gains or up to $3,000 more deducted against your adjusted gross income until the total amount is extinguished. You can also deduct transaction fees related to the cryptocurrency purchase, such as the percentage that an exchange takes whenever it processes a transaction as well as mining fees. Notably, however, investment-related fees may or may not be included in your cost basis for 2017 and prior. Tax professionals are not in agreement on this as relates to cryptocurrency. I believe that since the IRS is treating cryptocurrency as property, you can include it in your basis. Others argue that if you itemize your deductions, you can list any fees incurred for cryptocurrency trades in 2017 using Schedule A. In 2018, the rules are different. Whether you include this in your basis or not has real implications for 2018 into the future. Before you can calculate gains or losses, you must start by establishing your cost basis.

## 2. ESTABLISHING A COST BASIS FOR YOUR CAPITAL GAINS OR LOSSES

Your cost basis is how much you actually paid for a virtual currency when you purchased it, adjusted for any related costs. But when you are buying and selling coins as part of a portfolio, this can become more complicated than it seems. There are four commonly-accepted methods of figuring out your cryptocurrency cost basis: first-in-first-out ("FIFO"), last-in-first-out ("LIFO"), specific shares, or average cost. As a result, you may have the option to use the technique that works best for you.

1. Under FIFO, the coins you acquired first are also the first sold.

2. Similarly, under LIFO, the last coins acquired are the first ones sold.

3. The specific shares method is a bit more complicated. Under this method of establishing cost basis, the investor identifies the exact coins he or she is selling in order to establish the most favorable cost basis. This requires investors to keep detailed records and, if working with a broker, provide specific instructions regarding which coins are being sold.

4. Finally, the average cost method establishes a cost basis based on the average price of each coin in your wallet at the time you purchased them.

While this may all sound simple enough, things can get complicated very quickly once you actually start crunching the numbers. For example, assume that you made three Bitcoin purchases in 2017: you bought one coin in July for $2,500; a second coin in October for $5,000; and a third coin in December for $10,000. Imagine that you decided to sell one Bitcoin at the end of December when Bitcoin surged to $17,500. But which coin are you selling? The four different methods each have different answers.

1. Under the FIFO method, you are selling the first coin you purchased that year for $2,500. This creates a $15,000 short-term

capital gain that would be taxed according to your individual income tax bracket.

2. Under the LIFO method, you are selling the last coin you purchased at $10,000. Under this valuation method, your short-term capital gains are only $7,500, half of what it would be under FIFO.

3. Under the specific share method, you could select which coin you wanted to sell, which could differ for each person depending on their financial situation and goals.

4. Finally, under average cost valuation, you establish your cost basis at $5,833, which is the sum of all your coin purchases ($17,500) divided by the number of purchases you made (3). So, under average cost valu-

ation, your short-term capital gains would be $11,667.

Now, it may seem like the "obvious" choice to choose whatever option creates the lowest tax liability at the time. However, this is not necessarily the case. In fact, it may make more sense for you to pay higher capital gains this year if you anticipate that next year your income will bump you up into a higher tax bracket. If you plan to make more money year over year, you may want to claim as much income this year as you can, since in the following years your short-term cryptocurrency profits would be taxed at higher and higher rates.

Unfortunately for those who are used to self-preparing their taxes, calculating capital gains for cryptocurrency investments (and choosing which method to use) is anything but simple, especially if you've made multiple purchases, sales, or trades throughout the year. Even most do-it-yourself tax preparation software is not helpful when calculating capital gains, as these programs limit the amount of data you can import. For cryptocurrency traders, this can be a real shortfall. As a result, your best bet is to work with a licensed tax professional like a CPA, EA or tax attorney to figure out which accounting method will save you the most on your tax bill.

## 3. FOREIGN CRYPTOCURRENCY IN-VESTMENTS

There are numerous financial reporting and tax regulations affecting cryptocurrency held or purchased from a foreign broker or exchange. These rules are discussed extensively in the chapter of this book dealing with foreign asset reporting. The following discussion, however, focuses on the tax implications associated with buying or selling cryptocurrency assets through an offshore entity that you could potentially own.

There are several tax myths circulating around the internet regarding offshore companies being tax effective tax shelters. This myth is false. In fact, the IRS imposes passive foreign investment taxes on most income derived from non-U.S. companies, as well as offshore companies owned by Americans.

If more than half of an offshore company is owned by American citizens or permanent residents, it qualifies as a Controlled Foreign Corporation ("CFC"). The CFC rules were established to help curb tax evasion, which historically has been done by setting up offshore companies in low or zero-tax jurisdictions. CFC rules are defined and regulated differently in each jurisdiction, as the United States deals with the issue on a country-by-country basis through individual tax treaties. While the details vary substantially based on the location, operations, and other circumstances, a business formed in a foreign jurisdiction but owned predominantly by Americans is subject to U.S. taxes.

In general, foreign companies who conduct foreign business are entitled to an income tax deferment in the U.S. until the income is distributed as a dividend or income to U.S. citizens or permanent residents. CFCs, however, are not entitled to this deferment. Instead, Americans who receive passive income from offshore companies controlled by people inside the U.S. are subject to tax liability under Subpart F of the Internal Revenue Code. Subpart F taxes are levied based on ownership, not actual distribution of foreign passive income. As a result, Americans involved in CFCs can be taxed on the company's income in a given tax year even if the company did not distribute it back to its shareholders. In essence, the U.S. government is treating the foreign company like a domestic one for tax purposes because it is predominantly American owned. U.S. taxpayers must report their ownership in foreign companies using IRS Form 5471.

Some people who have set up offshore companies to buy and sell cryptocurrencies have also unwittingly exposed themselves to the IRS rules regarding Passive Foreign Investment Companies ("PFIC"). As defined by the IRS, a PFIC is any foreign company that takes 75 percent or more of its income in the form of passive income or if 50 percent or more of its assets could produce passive income. Anyone who receives income from a PFIC must pay taxes according to a complicated set of rules that the IRS has developed to deal with passive foreign income.

A company can avoid the extremely challenging tax environment of PFICs if it makes most of its income from a non-passive activity. However, offshore companies set up to buy and sell cryptocurrencies most often qualify as PFICs due to the fact that they generate passive income through coin investments. As a result, any U.S. taxpayer who receives income from these companies is subject to the PFIC tax rules.

The PFIC tax rules are famously complex. Under these rules, your tax liability depends upon how long you were involved in the operation, the distributions you received from the offshore company, and several other factors. PFIC investors must fill out IRS Form 8621 reporting the income they received from the offshore company, which is extremely lengthy and complicated. The IRS itself estimates that filling out an IRS Form 8621 could take over 40 hours – and that's for people who actually know what they're doing.

If you're involved in an offshore company that buys and sells cryptocurrencies, be sure to contact a qualified tax professional right away, if you haven't done so already. Dealing with PFIC and CFC taxes is a serious undertaking, but there are some possible upsides for people who know how to navigate these waters properly. If you are planning on buying or selling cryptocurrencies through an offshore account, be sure to work with a reliable licensed tax professional and attorney throughout the year in order to properly set yourself up for success. The costs of doing so generally start at $50,000 for the initial set-up in order for it to be done right.

# [4]

# CRYPTO DONATIONS, GIFTS & INHERITANCES

THE IRS PROVIDES TAX BREAKS for generous individuals who give away their cash or property. The same is true for cryptocurrencies. And just like with traditional currency, giving away your Bitcoin to a charity is treated differently than gifting it to a friend or leaving it to a grandchild in your will.

This chapter covers some of the basic things every investor should know about how the IRS treats charitable donations, gifts, and inheritances transferred in the form of cryptocurrencies.

## 1. CHARITABLE DONATIONS

Charitable organizations have to jump through a lot of hoops to receive 501(c)(3) tax-exempt status. If you give away your cryptocurrency to a qualified charity, you are generally entitled to an income tax deduction for the full fair-market value of the coin you donated.

More specifically, the IRS permits you to deduct the full fair-market value of the donation, plus fees associated with the transaction, up to 30% of your adjusted gross income. However, if you donate more than the 30% allowed annually, you can roll forward the unused deduction from the donation for up to five years.

So, for example, let's say you bought a bitcoin when it cost $1,000 and then donated it to your favorite 501(c)(3) certified charity when the price of bitcoin was at $13,000. Say your adjusted gross income that year was $21,000. In this case, you can deduct $7,000 from your income tax returns during that year, and roll forward an additional $6,000 deduction into the next year. Even better, as long as you waited to donate that coin for at least one year after buying it, you'll avoid paying capital gains tax on the $12,000 in profit you chose to forgo. If you hold a virtual currency for less than one year before donating, it still entitles you to a deduction, but you'll also be required to pay short-term capital gains.

Several popular charities now accept Bitcoin or certain altcoins. Depending on what causes you're passionate about, you can donate cryptocurrencies to 501(c)(3) organizations that help the homeless, encourage musical and artistic expression, promote peacekeeping and combat human rights abuses, or uplift individuals in developing countries. If you donate an amount of cryptocurrency worth more than $500, be sure to include a Form 8283 with your tax return. If you donate cryptocurrency worth more than $5,000, the IRS requires you to get an appraisal of the property and send it in with your tax returns, along with a written and signed acknowledgement from the charity that states the value of the coins you donated.

Charitable giving used to be one of the more straightforward aspects of tax law. However, the tax reform passed in 2018 has made things much more complicated. Because the new tax policy raises the standard deduction to $12,000 for individuals (and $24,000 for married couples) – almost double last year's rate – there is less of a financial incentive to itemize your deductions. In other words, if your itemized deductions don't add up to $12,000, you should would just take the standard deduction rather than the itemized deductions you may otherwise be entitled to. This can have a huge impact on charitable giving, which is one of the itemized deductions impacted by the new tax policy.

## 2. CRYPTOCURRENCY GIFTS

Since there are a lot of requirements for companies to qualify for 501(c)(3) tax-exempt status, you may want to give to an organization that has not officially qualified. Or perhaps you'd like to give a gift to friends or family members. These payments are not tax-deductible donations. Rather, they are gifts.

According to the IRS, any direct or indirect transfer to an individual is a gift if it's conveyed without the expectation of getting anything in return. You can exclude gifts from your tax liability up to a certain amount, but the rest is taxed. The recent Tax Cuts and Jobs Act increased the amount of gifts you can give without paying a gift tax. In 2017, you could exclude up to $14,000 ($28,000 for married couples filing jointly) in gifts per giftee, and any excess would have been subject to the 40% gift tax. In 2018, the gift exclusion was raised to $15,000 ($30,000 for married couples filing jointly). Since this amount is per giftee (whether an individual or organization), it means if you want to give $15,000 to your niece, $15,000 to your nephew, and $15,000 to a non-tax-exempt charity, none of those amounts will be subject to the gift tax.

The IRS also limits the amount of excludable gifts you can claim on your taxes for your entire lifetime. In 2017, individuals were not able to claim tax deductions for gifts or estate transfers of more than $5.49 million ($11 million for joint filers). The Tax Cuts and Jobs Act raised this to $10 million. As a result, individuals can give away cryptocurrency up to $15,000 per year, with additional amounts up to $10 million over his or her lifetime, without paying taxes out of the gifted amount in order to do so. This does, however, require a special tax form to be filed. Make sure you work with a knowledge-able tax professional if you fall into this cat-egory. This is all based on the value of the cryptocurrency at the time you gift it, even if it increases or decreases after that. It is very important to note that your cost basis for the gifted crypto passes on to the giftee so this is not a strategy to eliminate taxes, only defer them and pass on ownership of these assets to others.

## 3. INHERITANCES IN VIRTUAL CURRENCY

When someone passes away, his or her property is typically passed on to beneficiaries through an estate. The property is usually assessed at the fair-market value at the time of death or within six months thereafter. If you inherit cryptocurrency, document the fair-market value at the date of the decedent's death. If you sell your new crypto assets immediately, you must report the sale to the IRS, but you will not incur any capital gains.

If you purchase cryptocurrencies and then never sell them, but rather pass them along to your heirs, you can avoid paying capital gains on the assets while you're alive. This has tempted some beneficiaries into trying to game the system by transferring assets to the decedent shortly before he or she passes away with the intention of re-establishing the cost basis to fair-market value at the time of death. However, the IRS has seen pretty much every trick in the book at this point, and it will usually adjust the taxes back to the proper basis (and, of course, impose the fees and penalties it sees fit to ensure that you don't try to pull off a similar scheme in the future).

[5]

# CRYPTO IRA, 401K & LIFE INSURANCE

RETIREMENT ACCOUNTS and life insurance policies offer substantial tax breaks to the beneficiaries and account owners. But are these same tax benefits available to cryptocurrency investors using these financial vehicles? The following discussion highlights some of the potential benefits and pitfalls of retirement account and life insurance policies funded using cryptocurrencies.

## 1. CRYPTOCURRENCY RETIREMENT ACCOUNTS

It's an unfortunate fact that most Americans do not save up enough to retire comfortably. As a result, the government has created significant tax benefits for retirement accounts like IRAs and 401(k)s (the traditional or Roth versions) in order to encourage people to save more money. Some cryptocurrency investors have purchasing these assets though their retirement accounts to avoid or delay paying taxes on their virtual currency profits.

As discussed previously, the IRS treats Bitcoin and other cryptocurrencies as capital assets, much like stocks and bonds. As such, virtual currencies can be held in low-tax retirement accounts so long as they are not bought and sold in contravention of the prohibited transaction rules in Section 4975(c) of the Internal Revenue Code.

By holding your coins in a retirement account, you can take advantage of some significant tax benefits. All income and capital gains that accumulate from the crypto assets in your retirement account will be tax-deferred if you have a 401(k) or tax-free if you have a Roth IRA. The IRS disallows certain transactions that don't primarily benefit the retirement account. Prohibited transactions include those made between an IRA and "disqualified persons," as defined by Section 4975(e)(2) of the tax code. There are several scenarios that may give rise to a prohibited transaction, but they generally arise when the IRA transacts with the IRA account owner, the account owners' friends or family, or companies that the IRA account owner controls. To avoid prohibited transactions, most investors who purchase cryptocurrencies through a retirement fund do so through a Self-Directed IRA or Solo 401(k).

Self-Directed IRAs are managed by the account holder under the advice of an IRA custodian, who may require you to set up an

LLC to perform coin trades on your behalf. In these cases, the LLC would be fully owned by your IRA. Because the company qualifies as a disregarded entity, you would not be required to file a separate tax return. Although, some states impose special requirements on companies like this, so be sure to discuss your options with an attorney or CPA licensed in your area.

If you are self-employed or run a business with no full-time employees, then you may be able to set up a Solo 401(k) plan to buy cryptocurrencies for your retirement. In this case, you would serve as the trustee of the account. That means you have control over all of the funds in the account, and you can buy and sell cryptocurrencies through your 401(k) at will. Many investors, however, choose to form a special purpose LLC under the 401(k) plan to make transactions. Depending on your chosen cryptocurrency exchange, it may be more efficient to open a trading account using an LLC rather than your 401(k). Just as in the case of buying cryptocurrencies through your Self-Directed IRA, you have several options. As a result, it's a good idea to consult with a trained cryptocurrency accountant before shifting your cryptocurrency investments into your 401(k).

If you choose to invest in cryptocurrencies for your retirement, you should be aware that doing so often requires you to become the manager for your own investment account. IRS rules prevent you from borrowing from the account or otherwise profiting from it personally, just like any other professional investment advisor. Furthermore, total annual contributions to your IRAs cannot exceed a combined $5,500 if you're under age 50, and they're capped at $6,500 once you're older. So, even if you make it through all the hurdles of establishing and managing cryptocurrencies through your retirement account, this contribution limit may still leave you with plenty of virtual currency assets that you'll have to pay capital gains tax on if you want to invest more than the allowed limits.

There are also possibilities of rolling over other retirement accounts into a Self-Directed IRA or Solo 401(k) if you have accumulated capital in those accounts you'd like to diversify into cryptocurrency investments.

## 2. BUYING CRYPTOCURRENCY THROUGH A LIFE INSURANCE POLICY

Some investors use life insurance policies as investment funds in order to take advantage of some of the available tax breaks. Life insurance policies can have tax breaks similar to retirement accounts. For example, if you set up a private placement life insurance policy, hold it for a period of time, and then cash it out, you are entitled to tax deferral similar to a traditional IRA. You'll still pay capital gains tax, but you can defer your tax liability until a later date. There's no tax break, just a deferral.

When you use your life insurance policy as an investment vehicle, it's sometimes called a "wrapper" because you have essentially wrapped up your traditional investments in a different financial vehicle. However, the IRS tends to see scenarios like this as potential for tax evasion so the tax agency imposes certain fact-based standards to determine whether a wrapper should qualify for beneficial tax treatment when it is cashed out. Depending on the circumstances, the IRS could treat a life insurance wrapper as a modified endowment contract, which means most of the tax benefits of setting up the investment would be lost. Considering the fees and penalties many life insurance policy owners face when cashing out their policies, the financial risk of managing tax liability by buying cryptocurrencies through a life insurance policy is seldom worth the hassle.

But what if you never cash out your policy? If you hold virtual currencies in your life insurance policy until your death, it passes to your heirs. Like other property conveyed through life insurance, your heirs wouldn't have to pay taxes on any capital gains on the virtual currencies held in the policy. Sounds great, right? After all, isn't the whole point avoiding paying capital gains on your cryptocurrency earnings? Sure, except for one big problem: you'd be dead. What's the point of avoiding taxes if you're never able to spend the money you're trying to hide from the IRS? You may as well just consider giving the IRS its cut and enjoying life while you're still living it.

Leaving cryptocurrencies to your heirs sounds like a nice way to provide for your family after you're gone. However, life insurance policies are often not the most efficient means of achieving this goal. As a result, if you want to pass your virtual currency investments along to your heirs, it's best to do so through a more traditional estate plan developed with the assistance of an attorney.

[6]

# VIRTUAL CURRENCY LOSSES

THEFT AND FRAUD are major issues in the crypto market. In 2017, hackers made off with $500 million worth of cryptocurrency from Coincheck, a popular Bitcoin exchange. This theft shattered the previous world record, a staggering $400 million in Bitcoin stolen from the Mt. Gox exchange in 2014. And exchanges are not the only targets for cyber-criminals. By performing an extensive analysis of the 110 top-performing Initial Coin Offerings ("ICOs") over the past few years, researchers have found that more than 10 percent of the $3.7 billion raised by these new coins were stolen. This approximately $370 million loss is due to theft separate and apart from the market impact of fraudulent ICOs, which are a huge problem as well.

Unfortunately, the risk of loss by theft or fraud is astronomically higher for cryptocurrency investors than people who put their money into more traditional regulated securities. The following discussion highlights some tax implications raised when you sustain cryptocurrency losses, as well as some important tax distinctions between theft, fraud, and investment losses.

## 1. CRYPTOCURRENCY THEFT

Due to the inherent complexities of cyber-security, theft by hacking is an unfortunately has been a somewhat common occurrence in the crypto space. Most often, hackers have stolen digital coins by accessing an investor's private wallet using spyware, malware, or phishing techniques. Once they get your private key giving them access to the wallet, they simply empty the account.

If you've been the victim of a hack, it's important to know what options you have to recover at least some of your money. Even if cyber-crime investigators can't track down your virtual currency, the IRS may give you a tax break. While this cannot make you whole again, it can at least help offset your loss. In general, the IRS allows taxpayers to deduct losses from the theft of money or property, as long as they qualify.

For 2017, the IRS allows you to list any cryptocurrency you lost due to theft as an itemized deduction. The Tax Cuts and Jobs Act eliminates this deduction for 2018 forward. All examples in this Theft section of this chapter apply only to 2017 thefts. If computer hackers emptied your wallet or you invest in a new coin or project that turns out to be fraudulent, you may be able to recover some of your loss on your tax return. Theft losses are deductible in the year that they are discovered, and only if there is no reasonable chance of recovery. Under the tax code, "theft" covers all criminal means of appropriating another person's property, including swindling or false pretenses. This definition is tied to the criminal code in each taxpayer's particular jurisdiction, as theft for tax purposes requires criminal intent. Notably, theft is different from fraud, which is discussed below.

If your circumstances substantiate a theft loss, the amount of your deduction depends on whether the theft occurred in a for-profit transaction or an unrelated activity. For instance, if you paid for a car using cryptocurrency and then the seller disappeared, that would count as theft during a for-profit transaction. For profit transactions are deductible as itemized deduction. If, however, the coins were stolen in a hack while sitting in your wallet, may be what is considered an unrelated activity. That is, if your coins were stolen in some manner not connected to a transaction entered into for profit, you can only deduct losses that exceed 10% of your adjusted gross income, minus $100. Therefore, to calculate the deduction, start with the cost basis for the stolen cryptocurrency (what you paid for it, plus any fees incurred) and then subtract $100 and 10% of your adjusted gross income. What you're left with is the allowable deduction for the loss, which you report as an itemized deduction. Altogether, how much you can deduct depends on how much money you made in that particular year. Also, because the theft loss is an itemized deduction, if you take the standard deduction rather than itemizing you may see no tax benefits on the loss.

If you're among the many investors impacted by theft or fraud in the cryptocurrency market, consult with a tax professional right away. Whether you qualify for a theft or capital loss deduction can be a complex determination, and it's important to get everything straight before you claim a deduction for a cryptocurrency theft.

## 2. CRYPTOCURRENCY FRAUD

If you lost cryptocurrency in a dubious Initial Coin Offering, you may not be able to take the theft loss deduction due to the criminal intent requirement. In these cases, though, you may still be able to write off your costs as a capital loss. Even if the ICO turns out to have failed due to fraudulent activities of the company's officers and directors, a loss from stock tradable on the open market is almost always treated as a capital loss. You can only deduct up to $3,000 in capital losses per year with any amount above that carrying forward to future years, so your immediate financial recovery from failed ICO investments is limited.

Consumers who experience loss through hacking, fraud, or simply losing their private keys (thus losing access to their wallet and any currencies inside) can file a complaint with the Consumer Financial Protection Bureau ("CFPB"). Although it's never been tested in court, the CFPB may be able regulate virtual currency under the Electronic Fund Transfer Act ("EFTA") and the Federal Reserve's Regulation E. However, given the current political climate and a lack of case law to support CFPB action, there may be little that the agency can do to help you especially if it were hackers and not an actual project or exchange that defrauded you. As a result, vigilance is the best defense against potential fraud. Investors have already lost millions of dollars in fraudulent ICOs. In order to not become one of them, you must diligently research the company offering the coin in the first place.

## 3. CRYPTOCURRENCY INVESTMENT LOSSES

As discussed in other sections of this book, you can use capital losses to offset capital gains in your cryptocurrency investments. So, if you earned a good return on one coin but lost money on another, you can use the losses to offset your gains. However, the amount of losses you incurred on down cryptocurrency trades are only deductible up to $3,000 per year against your ordinary income. As a result, if your cryptocurrency portfolio is down across the board and you have sold enough of it to realize over $3,000 of losses, you won't be able to use the excess to offset your tax liability however can carry any additional amounts forward to offset gains in future years or claim the $3,000 deduction against your ordinary income each year until the total is used up if you had no other gains to offset in future years.

# [7]

# CRYPTOCURRENCY

# GAMBLING

INVESTORS WHO BOUGHT BITCOIN prior to mid-2017 are currently winning big. Some people in the gambling (or "gaming") community have already been using cryptocurrencies for many years. As a matter of fact, cryptocurrency gambling platforms manage dozens of the world's most active Bitcoin wallets. Experts estimated that as of 2017, up to 60% of all Bitcoin transactions – over 300 transactions per second – are performed for gambling-related purposes. One forward-thinking Irish company has even launched a Bitcoin Lottery that's on track to raise serious competition for established giants like the Powerball and Mega Millions.

However, just like the mafia of old Vegas, the IRS wants its cut after the fun and games are over. Anyone who wins money gaming or wagering must pay taxes. This includes everything from typical card games, sports books, and casinos, to racetrack bets, game show winnings, lotteries, and even Bingo.

## 1. CRYPTOCURRENCY GAMING: A GROWING INDUSTRY

Gambling sites give users the option to play using digital currency for several reasons. First of all, cryptocurrency transactions are very fast. They can be performed instantaneously in some cases, making them much more efficient and convenient than waiting days for a bank wire to clear. Secondly, every transaction made on the blockchain is publicly available. This makes all gambling payments clear and transparent, something that the gaming industry has come to value. And finally, cryptocurrency processing fees are usually much lower than traditional bank fees. In some cases, virtual transactions are almost free.

All in all, online gaming sites value cryptocurrencies for their high degree of security and liquidity. And online gamers themselves enjoy the ease and flexibility of using virtual currencies as well. Using crypto to play casino games or bet on sports can extend game time, since most coins are divisible down to several decimal places. This makes cryptocurrency suitable for micro-bets, so online gamers can enjoy the casual fun of gambling for hours and hours without losing their shirts. And there is substantial evidence that they are doing just that, as the average bet of an online gambler using Bitcoin is markedly lower than the online gaming industry average.

Blockchain technology within the gaming industry is advancing at a breakneck speed, and brick-and-mortar casinos are already working to introduce cryptocurrency-payable games on their floors. In fact, smart roulette tables fitted with QR scanners and touchscreens have already been developed. Soon, casino visitors will be making bets from their digital wallets just as often – if not more often – than they wager money from the wallets in their pockets and purses.

## 2. TAXES ON CRYPTOCURRENCY WINNINGS

The IRS has developed particular tax policies regarding wagering. Regardless of whether gamblers are paid out using traditional cash or cryptocurrency, or even with valuable prizes like vehicles and vacations, winning a bet or game creates taxable income. As a result, every gambler must be aware of the tax rules that apply to their particular circumstances.

Currently, the tax code requires gamers or casinos to withhold 24% of all gambling winnings over $5,000 (or 300 times the amount of the original bet) for remittance to the IRS for taxes. The IRS also requires a "backup" withholding of an additional 24% in some cases. Depending on where you play, your casino or gaming site may withhold the required taxes for you. When they do, they will issue you a Form W-2G, which is used to report winnings and withholdings to the IRS. However, you must report all gambling winnings to the IRS even if you don't receive a Form W-2G.

Typically, gambling wins and losses are reported separately. Gambling losses are disclosed on Schedule A as a miscellaneous deduction, whereas gambling winnings are reported on the Other Income line on a 1040. Most gambling income is reported on a Form W-2G, but you should use a Form 5754 to report any winnings you've made as part of a group, like a lottery pool, or for winnings you made on another person's behalf. Gamblers who itemize deductions may deduct their losses up to the amount of their winnings, but they are not subject to the 2% limits of other deductions.

As mentioned throughout this text, you must keep good records in case the IRS audits your tax return. This is just as true for gamblers as it is for every other taxpayer. Cryptocurrency gamers should keep detailed financial records of all gambling wins and losses to make sure they are able to substantiate information on key tax forms like Form W-2G and Form 5754. In fact, if you plan to

deduct gambling losses (or any other deduction for that matter) you must be able to produce evidence to the IRS that you are
entitled to it. The IRS recommends keeping
a gambling log, but this may not be necessary
if you play online. More often than not,
online casinos record your history, so your
taxable activity is easily accessed. The only
caveat would be if they cease operations, so
the burden will always remain on you. Regardless of the method you choose to track
your gambling activity, you will need to be
able to come up with proof of when you made
each win (or loss), what type of gambling activity was involved, the casino (online or
brick-and-mortar) you gambled at, and the

amounts of your wins and losses. Depending on your habits, this information can pile up quickly. As a result, working with a good accountant is a smart way to make sure you have all of your gaming records in order.

# [8]

# EXCHANGING VIRTUAL CURRENCY

CRYPTOCURRENCY WAS INVENTED to create an easier and more efficient means of exchanging money online. However, there are several different types of online financial transactions, and financial regulators treat each one differently. The following text explains the differences between coin-for-cash exchanges and coin-for-coin exchanges, including the major tax and regulatory issues raised by each type.

## 1. COIN-FOR-FIAT EXCHANGES

In nearly all U.S. jurisdictions, a business must be licensed as a money transmitter if it transmits funds from one person to another – including swapping cryptocurrencies for cash. Operating an unlicensed money transmission company is prohibited by the laws of 48 states and the District of Columbia, as well as the USA PATRIOT Act and the Bank Secrecy Act.

For the most part, federal money transmission is regulated by the Financial Crimes Enforcement Network ("FinCEN") within the Department of the Treasury. FinCEN's main goal is to deter money laundering and the financing of terrorism. As a result, it requires financial institutions, including money transmitters, to register with the agency and implement specific anti-money laundering programs.

Blockchain currencies themselves don't directly qualify as money transmitters because financial transactions are decentralized, not performed by a single person or company. However, according to FinCEN, certain actors in the cryptocurrency economy do fall under the definition of money transmitters that must comply with applicable regulations. Specifically, anyone engaged in the exchange of virtual currency for real currency, virtual currency, or other funds for business purposes must register with FinCEN and comply with the agency's anti-money laundering rules. Similarly, a person who issues a virtual currency for business purposes and has the authority to withdraw that currency from circulation must register and comply with all anti-money laundering regulations (this may seem to include ICO issuers themselves). Individuals and businesses who fall under FinCEN's jurisdiction are also likely to require state money transmitter licenses, but this depends on the laws of the particular state in which the exchange took place.

The Department of Justice has not been shy about prosecuting individuals who are transmitting cryptocurrencies without a license. Ripple (with the token symbol "XRP") rose to the upper echelon of the cryptocurrency market in 2017. But back in 2015, Ripple Inc. was facing a FinCEN enforcement action for operating as a currency exchange service without properly registering. Specifically, Ripple faced an enforcement action for facilitating transfers of virtual currencies in exchange for traditional money or other virtual currencies. The case was settled for fees and penalties, and Ripple was allowed to continue operating.

However, the agencies didn't stop there. The Feds aren't just going after the big fish in the cryptocurrency exchange world. Recently, the FBI and the Treasury Department have been prosecuting individuals for offering in-person cash exchanges for electronic transfers of Bitcoin. This type of service is relatively common, and several websites connect cryptocurrency investors with people willing to send them cryptocurrency for in-person cash payments. These people may be surprised to find out that this practice is most likely illegal.

In late 2017, the president of the largest regional chapter of the Association of Information Technology Professionals pled guilty to conducting an unlicensed money transmitting business. Why? For offering local sales of bitcoins through LocalBitcoins.com.

The defendant was caught by a sting operation run by undercover Treasury agents. The transaction he was charged for was one that many members of the crypto community are familiar with. Specifically, he advertised in-person cash-for-bitcoin exchanges through LocalBitcoins.com. On several occasions, he met up with undercover agents posing as customers where he accepted cash payments for electronic bitcoin transfers. Because he was not licensed as a money transmitter in his state or with the Financial Crimes Enforcement Network, each transaction violated state and federal law. As of the writing of this book, this particular defendant had been adjudicated guilty but had not yet been sentenced. But, he is facing up to five years in federal prison, even though his earnings from the unlicensed exchange was only $2,122 in fees he collected during the meet-ups. While it surely seemed like good money at the time, he no doubt regrets his decision in retrospect. What many familiar with this case do not realize is that the case was initiated by the IRS's Criminal Investigation Division.

As the individual in the above example learned the hard way, ignoring financial regulations can expose you to massive legal and financial liability. The federal government has never been shy about flexing its enforcement muscle, and you don't want to find yourself on the wrong end of a federal case. As a result, complying with all federal financial laws is critical to making sure your wallet – and your freedom – is protected from the long arm of the law.

## 2. COIN-FOR-COIN EXCHANGES

Exchanging one cryptocurrency for another in a coin-for-coin trade is an extremely common occurrence in the virtual currency investment community. In fact, some high-volume traders do so every day. Coin-for-coin swaps do not raise the same money transmission issues discussed above because, unlike cash, the federal government does not define these assets as "currency." As a result, most of what you need to know about the tax implications of coin-for-coin trades is addressed in the section discussing cryptocurrency investments. However, there are two particular types of coin-for-coin exchanges that bear specific mention: wash sales and like-kind exchanges.

## A. WASH SALES

Cryptocurrency markets have been slipping in the beginning of 2018, and some investors may be seeing losses as a result. But, if you sell your cryptocurrency holdings for a loss today, and then re-purchase them at near the same price tomorrow, can you claim a deduction without running afoul of the wash sale rule? The IRS has remained mum on the issue, but most experts agree that you can indeed.

An investor is entitled to a tax deduction for any capital loss he or she accumulates in a given year. Typically, selling a stock or security at a lower price than you bought it for qualifies as a capital loss – unless you repurchase the same stock or security within 30 days. These sorts of sales are considered wash sales, and they are excluded from the capital loss deduction allowance.

Wash sales are banned by the tax code because the loss only exists on the books, not in the investor's portfolio. Consider an example: an investor buys 100 shares of stock at $10 each, and then sells them when the market drops and the stock price slips to $5. At first glance, this seems to create a capital loss of $500 that the investor can deduct from her taxes. However, assume the investor repurchases 100 shares of the same stock a few days later when the market is still down, paying $5 per share. This puts the investor in the same position that she was in before the sale that caused the $500 capital loss. To prevent investors from using down markets to accumulate tax deductions on money they didn't actually use, the IRS does not allow any deduction in these circumstances.

The wash sale rule exists to prevent investors from getting tax breaks on artificial losses. However, under the tax code, the prohibition only applies to "any sale or other disposition of shares of stock or securities." So, on its face, the wash sale rule only applies to common stocks, options, and convertible securities. This means that as it stands, Section 1091 does not apply to Bitcoin or other cryptocurrencies because the IRS treats these assets as property. The IRS's stance on this matter could certainly change in future cases or guidance issued. Property assets do not fall within the strict statutory prohibition on wash sales of stock or securities. Because the wash sale rule does not apply based on the express language of the statute, crypto investors can probably claim capital losses from coins they sold and repurchased within 30 days.

Overall, most tax experts concur that you can perform wash sales in cryptocurrency trades without running afoul of the tax code. However, this may not be the case for long. Section 1091 does allow the IRS to expand the "stock or securities" that trigger the wash sale rule. If the IRS passes a regulation clarifying that Bitcoin and other cryptocurrencies do fall under the jurisdiction of Section 1091, wash sales may be disallowed forever.

## B. LIKE-KIND EXCHANGES

The IRS characterization of cryptocurrency as property rather than currency or securities has raised several highly debated tax issues. Among the most relevant of these novel questions is whether swapping one coin for another qualifies as a like-kind exchange under Section 1031 of the tax code.

A 1031 like-kind exchange is a very specific type of transaction involving a trade of one kind of business or investment asset for another, largely identical, asset. For example, if you own a furniture store and decide to exchange an orange sectional couch that your customers detest for a more fashionable leather pull-out, you may not be required to pay taxes on the exchange. So long as the two pieces of furniture are of the same nature of character you can defer paying taxes on the swap under Section 1031. Rather, in this case your tax liability arises when you sell the pull-out couch.

In 2018, the Tax Cuts and Jobs Act (TCJA) made a change to what type of assets were allowed to be utilized in like-kind ex-changes to limit them only to real estate. Some cryptocurrency investors have argued that they should be entitled to Section 1031 tax benefits for 2017 and prior because all virtual currencies have the same basic nature and character. However, this commonly-circulated tax myth does not accurately reflect how the majority of tax practitioners agree that cryptocurrencies must be treated in regards to 1031 exchanges. Dispensing of your virtual currencies in exchange for anything of value is a taxable event. This is true whether you sell your coin for U.S. dollars, trade them for other virtual currency, or spend them on retail goods and services. It is highly likely that coin-for-coin trades do not qualify as a swap of property-for-property as contemplated by Section 1031. Rather, these transactions are treated as a sale followed immediately by a purchase.

While the 1031 like-kind exchange exemption has raised a lot of crypto tax questions for years prior to 2017, changes to the rule created by the 2018 federal tax reform bill has put the issue to rest for all future tax filings. The Tax Cuts and Jobs Act expressly limited the application of Section 1031 to transactions involving real estate only, closing the like-kind exchange loophole question entirely as least as it pertains to virtual currency swaps.

For cryptocurrency investors who relied on the like-kind exemption to avoid paying taxes prior to 2018, they may want to get ready for a fight. The IRS has not released specific guidance on the issue of 1031 exchanges of cryptocurrency prior to 2018, which has given some investors a glimmer of hope that all coin-for-coin swaps prior to 2018 were tax-free. Given the breadth of the IRS's cryptocurrency tax policy, it is unlikely to agree. As a result, this issue will almost certainly be litigated in federal court. Unless you're willing and ready to pay substantial legal fees to defend your position regarding coin-for-coin swaps qualifying as like-kind exchanges, it's more prudent to simply pay the taxes due on all coin-to-coin trades.

# [9]

# CRYPTOCURRENCY TAX REPORTING

DEPENDING ON YOUR INVESTMENT AC-
TIVITY, you may be subject to onerous financial reporting obligations. The IRS imposes
tax reporting requirements on cryptocurrency investors, which are detailed below. However, your cryptocurrency reporting
requirements may not end with your tax re-

turn. Both the Treasury Department and the IRS impose registration and reporting requirements on most people who hold a foreign bank account or assets, which likely includes cryptocurrencies held on foreign exchanges.

The IRS requires you to report all income on your annual or quarterly tax returns, and this included money you made trading virtual currencies. Furthermore, the IRS expects you to maintain supporting financial documentation available for review upon demand. If you made more than 200 transactions totaling more than $20,000, your exchange should report your activity by issuing a Form 1099-K to you and the IRS. However, since

many cryptocurrency exchanges remain un-
regulated, so far for tax year 2017, only
Coinbase has issued 1099-Ks to qualifying
investors.

However, the 1099-K that some Coinbase
customers received is only the beginning of
the story. Your Form 1099-K shows all of the
transactions that passed through Coinbase in
a given calendar year, but it does not establish your cost basis. You must still report
your cost basis for each trade to the IRS using a Form 8949.

Your Form 1099-K gross receipts do not
need to match up exactly with your Form
8949, but you will need to substantiate any

differences between them. After all, the whole purpose of this exercise is to catch potential tax evaders. If you believe the 1099-K form you received from Coinbase is inaccurate, contact the exchange immediately. The form will need to be amended and re-submitted to the IRS if it has any errors, and this may impact your tax filing this year.

Many investors have used the helpful online tools available at Bitcoin.tax or Cointracking.info to generate the forms they need to accurately report their cryptocurrency gains to the IRS. You can use this tool to create the exports to create Form 8949 from your account history data downloaded from your exchange. That 8949 then flows into the Schedule D of your return. The IRS will look at your Form 8949 and Form 1099-K together to assess whether you are accurately reporting your cryptocurrency activity on your tax returns. All in all, if you've received a 1099-K from Coinbase, be sure to work with a qualified tax professional to make sure that your return is accurate and complete. It's a good idea for any cryptocurrency investor to work with an accountant or qualified tax expert when filing taxes this year.

When you submit your tax returns, you swear to the accuracy of the documents under penalty of perjury. So, if you get questions from the IRS or requests for documentation, you will need to be prepared!

If you haven't kept meticulous notes about every trade you performed last year, don't worry; many exchanges let you export your account activity to an Excel spreadsheet or PDF. But be forewarned – just looking at these documents may just make your head spin. And when it comes to trying to figure out what you need to calculate your cryptocurrency taxes from your raw account data, you may start getting waking nightmares of high-school algebra class.

Just to make things more complicated for cryptocurrency investors than they already are, the recent federal tax reform changed the rules of the game for many cryptocurrency investors. For example, starting in 2018, you can no longer include cryptocurrency-related transaction fees in your itemized deductions on your personal income tax return. (This deduction is still allowed for businesses.) The 2018 tax reforms also change the capital gains tax rates, which may greatly impact your investment decisions. Holding onto your cryptocurrency assets for another few months or not may save you – or cost you – thousands of dollars on your federal income tax returns.

The IRS expects you to be able to prove any income and deductions you claim on your tax returns, so smart investors are prepared with good documentation. Your accountant can help you prepare any records you may need to submit to the IRS. However, unlike other types of investments, much of the required documentation is only available online. If your accountant isn't sure about what records the IRS will require from you regarding your virtual currency investments, contact a trained cryptocurrency tax expert.

# [10]

# FOREIGN ASSET REPORTING

IN THE VIRTUAL WORLD, nationality is all but meaningless. The internet exists without meaningful political boundaries. At any moment in time, you can connect with a person from halfway around the world, no passport required.

However, the same cannot be said for the enforcement of laws. When it comes to financial activity, national citizenship should be a critical consideration. That's because the federal government imposes special reporting requirements on anyone holding foreign financial assets or accounts.

The IRS requires any American citizen residing in the U.S. holding foreign assets that were traded in excess of $50,000 ($100,000 if married filing jointly) on the last day of the tax year or more than $75,000 ($150,000) at any time that year to file a Form 8938, Statement of Specified Foreign Financial Assets or "FATCA" form. This minimum filing threshold increases to $200,000 ($400,000 for joint filers) and $300,000 ($600,000 for joint filers), respectively, of foreign assets for

American individuals who live abroad for more than 330 days out of the year.

While the issue has not been put to rest by a direct IRS policy statement or federal court case, cryptocurrency purchased from a foreign broker or exchange may likely qualify as foreign financial assets that trigger FATCA reporting requirements. The tax code defines foreign assets as any financial instrument or investment contract made between an American and a foreign counterparty. The definitions applied to these concepts are very broad. As a result, whenever an American purchases cryptocurrency on a foreign exchange – the assets are likely subject to

FATCA requirements if they exceed the minimum reporting threshold.

The FATCA form – so named for the Foreign Account Tax Compliance Act – is a component of the U.S. law enforcement regime specifically aimed at preventing and punishing tax evasion and money laundering. Consistent with this law enforcement purpose, the IRS imposes serious penalties on anyone who holds substantial foreign assets but fails to submit a FATCA form. The IRS can impose a penalty of $10,000 for failure to file, and an additional penalty of $50,000 for any continued failure to file after the IRS notifies you of the requirement. If your non-disclosed assets impact your tax liability, you may be subject to an additional 40 percent penalty for understating your taxes. Each of these penalties can accrue annually for up to six years.

If you've made large cryptocurrency asset purchases from exchanges or brokers that you cannot verify are American, you should to file a Form 8938 with your tax return. Further, assets held in foreign bank accounts are subject to additional reporting and registration requirements under FinCEN rules. This is because offshore bank accounts have been used to commit financial crimes, including tax evasion and money laundering by terrorists and organized crime.

The tax code requires you to report income to the IRS from all sources, worldwide. So, even money you made offshore must be reported on your tax return. Additionally, if your international dealings extend to any interest in a foreign bank or financial account, you must report it separately using Schedule B. This reporting requirement extends to money made on foreign cryptocurrency exchanges, which host wallets for millions of U.S.-based investors.

The IRS isn't the only thing you have to worry about if you're holding money in overseas banks or cryptocurrency exchanges. The Treasury Department requires every American who has $10,000 or more in a foreign account to file a Report of Foreign Bank and Financial Accounts ("FBAR"). This is true even if you don't maintain $10,000 in the account at all times, as the FBAR reporting requirements are triggered the total of all of your accounts exceed $10,000 in value at any time during the year. This is what the FinCEN website says: "A United States person that has a financial interest in or signature authority over foreign financial accounts must file an FBAR if the aggregate value of the foreign financial accounts exceeds $10,000 at any time during the calendar year." Although it may not be entirely clear from the quotation, the way the $10,000 is counted is not intuitive. The maximum value at any time during the calendar year for an account is added to the maximum value of all of your other accounts at any time during the calendar year to come up with the aggregate total that is used for these purposes. Consider this illustration. Someone has $7,000 in foreign crypto exchange Binance and during year they transfer all of the $7,000 to foreign crypto exchange Bitfinex. Someone might think to themselves, "Since I only had $7,000 I am not subject to these FinCEN reporting requirements." But that would be wrong. In this example, FinCEN adds the maximum in Binance to the maximum in Bitfinex to get an aggregate total of $14,000. This illustrates a problem of attempting to do your own taxes and FBAR filings. Your personal knowledge and assumptions may get in the way of properly complying with these tax regulations. Another example to keep in mind relates to the actual asset value of your cryptocurrency even if you didn't sell it. Imagine you bought a Bitcoin using Kucoin in October 2017 when the price was around $5000. Because it surged well over $19,000 before the end of the year, FBAR reporting requirements would be triggered, even if you didn't sell it until January 2018 when the price of Bitcoin went back down.

In the past, people could open Swiss bank accounts without even providing a legal name. Due to the nation's strict privacy laws, holders of Swiss bank accounts could stay nearly anonymous – making them attractive for people trying to hide money from the IRS. However, these days are long gone. Swiss banks have spent much of the last decade trying to shake their poor reputations, which means they're naming names to the IRS. Financial giants like Credit Suisse and UBS were caught up in IRS sweeps looking to nail tax evaders, and they're singing like canaries. As federal regulators continue to bring down the hammer on tax evaders and other financial criminals in the crypto space, it's more likely than not that foreign exchanges will follow the example of the Swiss banks, in fact, most of their Terms of Service state that they will provide details on account holders upon any government requests. Rather than spending the time and money fighting off the U.S. Department of Justice, theses exchanges are much more likely to simply turn over user data on U.S.-based accounts.

It's important to note that neither the IRS nor FinCEN have issued official policy statements regarding FBAR reporting requirements as they pertain to cryptocurrency specifically. However, there is some precedent in online gaming. In 2014, a federal court held that money kept in online gambling accounts were in fact subject to FBAR reporting requirements. As a result, it's not too much of a stretch of the imagination to believe that the courts would view assets held in foreign virtual currency exchanges the same way.

The penalties for keeping a foreign bank account without informing the IRS or Treasury Department are staggering. Not filing an FBAR is punishable by up to $10,000 in penalties per violation – and that's for an accidental omission. If the agency finds the FBAR non-reporting to be intentional, these penalties jump to $100,000 or 50% of the foreign-held account balance, whichever is greater. There is no statute of limitations on civil tax fraud, so the IRS can come after you for as many back taxes – plus interest, fees, and penalties – as they can find. Plus, the IRS can tack on a 20% inaccuracy penalty in these cases and a 75% civil fraud penalty applies in extreme cases.

Keeping money in a foreign account and not reporting it to the IRS or Treasury Department can cost you a lot of money, and it only gets worse from there. Both agencies can criminally prosecute you for not properly reporting a foreign account. Filing a false return is a felony, so hiding foreign-held assets from the IRS can earn you up to five years in federal prison. The IRS has broad authority to collect what it's due; the agency can bring a criminal tax evasion case against you based on tax returns filed up to six years in the past. As a result, tax evasion is the type of risky behavior that can catch up with you several years after you've decided to walk the straight and narrow. The Treasury Department, through the Department of Justice, can also charge you with a crime for failing to file an FBAR. These cases include financial penalties of up to $500,000 as well as a possible ten-year prison term.

If you are concerned about your potential liability for failing to properly report your foreign cryptocurrency exchange activity, consult a CPA or attorney right away.

# [11]

# CRYPTOCURRENCY TAX

# ENFORCEMENT

THE IRS CAST A WIDE NET when it Every year, the government loses over $300 billion in unreported income – more than half of total nationwide tax evasion costs. Because of this, the IRS is always keeping an eye out for people who appear to have more income than they are reporting on their tax returns. This is especially true for the cryptocurrency community, which has already been the target of IRS enforcement actions aimed at shaking out tax evaders.

In over 150 years of operation, the Internal Revenue Service has seen every trick in the book that investors use to avoid tax liability. The takeaway from this section should be to NEVER MISREPORT YOUR CRYPTOCURRENCY INCOME. To reinforce this point, the following discussion highlights the IRS' current policies regarding enforcement of cryptocurrency tax rules, as well as the likely fees and penalties imposed on inaccurate reporting. Keep in mind, the blockchain is a public ledger and will ultimately be the easiest set of audits the IRS ever has to undertake utilizing its Automated Underreporter (AUR) audit system.

## 1. ENFORCEMENT ACTIONS

The Internal Revenue Service has taken its time figuring out how it will deal with virtual currency investments. In the meantime, Bitcoin and the thousands of alternative coins that followed have made early investors millions of dollars. Until recently, these investors have largely evaded paying taxes. The IRS, however, isn't letting anyone off the hook.

Now that cryptocurrencies have made their way into mainstream finance, the IRS is aggressively pursuing investors who have not paid their taxes. Regardless of whether it's because of intentional evasion or just confusion, the IRS knows that cryptocurrency income is being under-reported, and the agency is hot on the heels of just about every major cryptocurrency investor. Avoiding tax liability altogether is not possible for most virtual currency holders, and attempting to evade federal taxes can land you in real trouble.

The IRS originally issued guidance that cryptocurrencies treated as taxable assets in 2014, and in 2016, the agency filed suit in federal court seeking retroactive tax assessments against millions of virtual currency holders. In the suit, the agency reported that it received only about 800 returns reporting Bitcoin-related gains or losses in 2015. Given that Coinbase currently had around 6 million users back then (currently over 15 million users), the IRS has argued that most crypto-currency holders are under-reporting their investment income.

In late November of 2017, Coinbase was ordered by a Federal Court to hand over information on thousands of user accounts. Pursuant to the order, the largest Bitcoin exchange in the United States supplied the name, birthday, address, and taxpayer ID and transaction histories of 14,000 user accounts to the IRS.

Why is the IRS interested in this information? Because it has substantial evidence that cryptocurrency investors have been making capital gains on their virtual currency trades and not properly reporting them on their tax returns. These 14,000 account holders can now expect an IRS enforcement action any day now as the data was turned over to the IRS in March of 2018. Those enforcement actions will include fees and penalties for failing to properly report their cryptocurrency income on their tax returns. At the end of it all, most of them will be wishing they had just paid their taxes properly in the first place.

Only 14,000 Coinbase user accounts are impacted by the recent lawsuit – so far. The IRS initially requested information on all Coinbase customers, and Coinbase was able to negotiate to limit the scope of the order to only a fraction of its active users. However, the initial IRS demand foreshadows greater enforcement across the board. The 14,000 Coinbase users now known to the IRS are the big fish – those who had Bitcoin investments valued at $20,000 or more at any time between 2013 and 2015. After this early win in enforcing tax laws against cryptocurrency investors, however, the IRS shows no sign of slowing down. It's only a matter of time until the IRS is able to acquire and comb through enough cryptocurrency transaction data to land lots of cryptocurrency investors in a whole mess of trouble. Ultimately, it's best to simply accept the inevitable and report your cryptocurrency income immediately, even retroactively if you failed to do so in the past. Not doing so could result in substantial fines, penalties, and even time behind bars.

## 2. PENALTIES FOR INACCURATE TAX REPORTING

You are required to report all cryptocurrency trading income – no matter how big or small – on your tax returns as capital gains. If you make crypto income and you don't include it on your tax return, you could get hit with serious fees and penalties. Not only will you have to pay taxes on the income you failed to report, you may be subject to additional taxes if the income discovered bumps you into the next tax bracket or affects your deductions. Even worse, the IRS can impose an inaccuracy penalty of up to 20% of the amount you underpaid. That's on top of a 5% monthly late fee on the amount you owe, up to 25%. Failing to file a return altogether in a year during which you owe taxes can land you a fine of up to $25,000 and a year in federal prison.

By way of example, say you failed to report $1000 in capital gains that you profited on Bitcoin in 2017. If you don't report it on your tax return this year, the IRS can come after you for the taxes you owe on this money as well as an extra $200 penalty for inaccurate reporting, plus up to $250 if you're more than five months late coming up with the payment. So, rather than paying the $1,000 you owed in capital gains tax if you had reported accurately and on-time, you could have to pay up to $1,450 in taxes on that $1,000 capital gain if the IRS discovers your omission. And this is just the penalty for unintentional underreporting. Intentional tax evasion is much more serious.

If the IRS decides that you intentionally attempted to evade taxes, it could decide to handle your case as a criminal matter. A willful attempt at tax evasion is punishable by a fine of up to $100,000 and up to five years in prison. It's possible that a careless cryptocurrency investor could face felony tax evasion charges for failing to properly report virtual currency income on their tax returns. If you filed inaccurate returns in the past, be sure to correct them immediately using a Form 1040X.

Evading federal taxes is a dangerous game. It can be tempting, sure – especially because it seems like everyone else in the cryptocurrency community is doing it. However, IRS penalties for tax evasion are severe. The IRS has considerable authority to collect the money it's owed. It can make you forfeit your tax refund, garnish wages and seize property to make up for the difference. The IRS can also revoke your passport or even file charges against you in federal court if you fail to pay your taxes. In the most extreme cases, crypto tax evaders may even land in federal prison. The IRS tends to be more sympathetic to individuals and businesses who file their taxes or correct for inaccurate reporting before the agency discovers the error for itself. Given that the IRS employs almost 3,000 special agents whose sole purpose is to uncover tax evasion, the risk of being found out is higher than it seems. The agency even has a whistleblower program which gives an informant a percentage of taxes they help the IRS recover. So, all it takes is one angry ex or irritated roommate to get you on the wrong side of the IRS. Evading taxes can cost you big in the long run, so it's much better to play it safe and properly report your crypto income on your federal tax return this year.

Generally, the IRS can audit you for errors in your tax returns from up to three years back. However, if you omitted more than 25 percent of your income on a prior tax return, an IRS audit can reach as far as six years into the past. And there is no time limit imposed on fraud, so any willful misrepresentation of your reported income is prosecutable forever. At the very least, this means you may get a question or a demand for proof from the IRS for a tax decision you barely even remember. As a result, keeping good records is critical. A good accountant can help keep your files in order for IRS inspection.

## 3. ANONYMITY WILL NOT PROTECT YOU

Everyone is entitled to the benefit of the doubt, but some cryptocurrency investors may be evading taxes because they believe they can never be detected due to the anonymous nature of must cryptocurrency transactions. However, despite popular belief, Bitcoin and most other cryptocurrencies are not completely anonymous. Save for a few coins specifically designed to provide greater privacy protections, the blockchain displays the public keys of all users who send or receive coin. The time, amount, and public keys involved in just about every coin transaction are publicly available. While users legal name, social security number, physical address, and other information required by traditional banks and financial service providers are not publicly posted, the identifying information publicly disclosed on the blockchain is often enough to mount an investigation especially when combined with the Know Your Customer (KYC) data that the IRS is seeking from the exchanges which is not anticipated to stop with the initial blow to Coinbase.

Public wallet keys are not linked to a person's actual identity. However, this link can be built by good old-fashioned research and investigation. And once a public key is linked with a person's identity, investigators can see every coin transaction the individual has ever made. Many people have put misplaced faith in the anonymity of blockchain currencies, to their great detriment. For example, the highly publicized arrest and conviction of Ross Ulbricht, founder of online black marketplace "the Silk Road," was partially based upon the government's ability to trace cryptocurrency payments sent from the Silk Road to Ulbricht's personal wallets. And in a fascinating twist to the Silk Road saga, two federal agents who had worked on Ulbricht's case were actually arrested themselves for allegedly stealing bitcoins from the Silk Road. Their activities were uncovered after analysis of the blockchain showed deposits to public keys linked to the agents' identities.

In short, cryptocurrencies are pseudo-anonymous and they do afford critical privacy protections. However, the typical degree of anonymity afforded in blockchain transactions is not sufficient to shield you from a federal investigation. If you believe your past cryptocurrency activity may expose you to criminal liability, speak with a tax attorney right away.

# CONCLUSION

Equipped with this information and the help of a trained tax professional, you'll be able to accurately file your cryptocurrency taxes (even if it takes a little extra time to sort it all out). Remember to keep thorough records and pay your taxes on time, and you shouldn't run afoul of the IRS. I implore you to think of it this way, if you are owing taxes, that means you are making money and that... is a good thing.

# ABOUT

# CRYPTOTAXPREP.COM

Recognized by Entrepreneur Magazine, Forbes, Crypto News Network, the New York Times, Bitcoinist, CoinCentral, CryptoCoin-Trader, Accounting Today and other leading publications, CryptoTaxPrep.com by Happy Tax is the industry leader in cryptocurrency bookkeeping, reconciliations, accounting, CPA tax preparation, foreign asset reporting and advisory. All tax returns come with 100% free audit assistance and are guaranteed to be accurate. Affiliate and content partnerships are available.

# DISCLAIMER

All of the information published herein is presented in good faith and believed to be correct, but it is general in nature and should not be taken as specific tax advice in any circumstances. The information contained herein may not be applicable to your particular situation, nor may it be suitable for your facts and circumstances. The information contained herein is not exhaustive in scope or particular to any single individuals' specific circumstances or needs.

As a result, the information contained herein may require consideration of other matters, facts, or circumstances not mentioned in this book.

Neither the authors nor the publishers assume any duty or obligation to inform any person of changes in tax law or any other factors that could possibly impact the information contained in this book.

# IRS Circular 230

# Disclosure

tax professional regarding their particular circumstances.

Neither the authors nor publishers of this text make any claims, promises, or warranties about the accuracy of the information provided in this book. Tax advice cannot be provided on a general basis. Rather, it must be tailored to each individual's particular circumstances by his or her chosen tax professional. Everything published in this text is the author's general opinion and shall not be treated as concrete fact.

# About The Author

Named as "One To Watch" by Accounting Today's 2017 Top 100 Most Influential People in Accounting, Happy Tax and Crypto-TaxPrep.com CEO Mario Costanz has overcome tragedy and obstacles in his 20+ year entrepreneurial career to become of the most successful entrepreneurs ever to enter the tax preparation industry. Mario excels as an Author, Leader, Speaker, Inventor and Visionary.

# CONTACT INFORMATION

Connect: https://linkedin.com/in/mariocostanz

Follow: https://facebook.com/mariocostanz

Email: Mario@TheProblemSolver.com

# THANK YOU

Thank you for taking the time to read my book. Without your support and the support of our many team members who assisted to make this possible, this book or my career would not be a reality. If I can ever be of assistance to you, please do not hesitate to reach out on social media or by email. If you enjoyed this book, please place a review for it on Amazon.

I'll leave you with my motto that has served me well over the years: There Is Always a Way!